Manual for Electricians

Volume 1

Basic Fault Finding Techniques
Motor Testing and Connections
Preventative Maintenance
Safety of Machinery
Circuit Design Methods
Examples of Electrical Circuits
Safe Working Practices

Author (Francois Korf)

Foreword

This manual is set as a guideline for fully qualified electricians. It is not to be used as a substitute for standards and regulations. As an electrician it is your responsibility to adhere to your local regulations and to always observe safe working practices.

I have written this manual as the sum of my personal experiences. It is to aid you in the fault finding of electrical circuits; to present you with best practices in preventative maintenance; to raise awareness of machine safety; and to build a foundation of knowledge in basic electrical circuits.

Continue to strive to be the best in your trade. Always deliver work to the highest possible standard. Take pride in the work you deliver.

Legal Notice(s) and Requirements

Whilst every care has been taken to provide true and just information in this manual, the author cannot be held responsible for any loss or damage caused by using all or any of the information. It is the responsibility of the person using this manual to ensure all work performed conforms to any legislative requirements and that all relevant safety procedures are followed prior to commencement of any work.

Copyright and Intellectual Property

ISBN 978-1-4717-9672-2

About the Author

Francois (Frans) Korf is a trade qualified electrician. He is currently living and working in Australia. Frans started his trade as an apprentice in the mining industry of South Africa.

Due to good fortune he has worked within a diverse range of industries and across multiple disciplines covering domestic, commercial and industrial installations. This has gained him invaluable experiences along the way.

For many years Frans has worked hands on in the electrical trade, managed electrical departments and maintenance departments (overseeing all trades). More recently he has been working as project manager implementing large industrial projects, and driving automation in a production environment.

His highest formal qualifications are shown here:

National Technical Certificate Level 5

Advanced Diploma of Project Management

Diploma of Business Management

Certificate IV in Training and Assessment

Index

Chapter 1

Basic Fault Finding Techniques

When you approach any installation to find a fault on it, it is more than just looking for a fault. In fact, fault finding is an art. You must force yourself to think in a logical manner and have a clear starting point from which you work through to the end. Once you start "jumping" from one place to the next, you are going to waste a lot of time. Saying this, let us have a look at the methodical approach to fault finding.

Key Elements to Fault Finding

In order for you to be able to walk up to a machine and locate the fault, you must have a good understanding of Electrical Circuitry. You may at times be fortunate and have a schematic electrical drawing and at times there will be nothing. If you do not understand an electrical circuit, you will be at a major disadvantage.

So let's consider the electrical circuit first. Apologies to the more experienced readers but we need to cater for our entire audience so please bear with me. I will just touch on the basics to give you a brief idea of what we will be talking about.

It is important to note the difference between the control and power circuits. When we fault find a machine, our main focus will be on the control circuit. This is the circuit that has all the stop and start buttons, limit switches, pressure switches, flow switches and the like connected to it. This will originate from a power supply and follows the control path through all the various switches up to either a relay or a contactor. Once the contactor receives power to the coil, it will energize and provide power to the motor or other device connected to it. If you follow this you can already see what our pattern will be to find the fault on the circuit right? Now let's break it down into a series of steps I would follow when arriving at a breakdown.

Step 1

Familiarize yourself with the operational surroundings.

Always inform the manager in that area that you are there to work on the equipment.

Ensure the equipment is isolated, locked and tagged out to prevent someone turning the power on while you are working on it.

Talk to the operator! The operator can tell you exactly what happened when the machine stopped, explain to you how the machine works if you are not familiar with it and maybe also tell you what normally goes wrong on the machine. These guys are your best friends at a breakdown! They tend to watch the sparky working on the machine and even though they may not fully understand what he/she is doing, they can point you to the spot that was worked on the last time. This is one of my Shortcut Rules, failing that; I will proceed to my more structured approach to troubleshoot the circuit.

An extremely important point to remember here is that you may have multiple supplies in the cabinet. One for the power circuit, and another for the control circuit. Please make sure you follow the correct testing procedure before touching anything. Your voltage tester is the most important item in your toolkit, **USE IT**!

Step 2

Check for any visible damage to the machine. This sort of damage may have caused a limit to shift etc. Ensure all the guarding on the machine is in place as these are generally connected to safety switches that will shut the machine down when not safe to operate.

This is also where you would have a quick look at the safety relay which has indication lights on it to indicate ready state or fault state. If there is no fault, proceed to the next step. However, if you have a fault indication you are almost there. Check each safety switch for operation and repair the faulty unit. Please NEVER bypass a safety switch!

If you have found the problem and repaired it, move to step 5. If not, continue on to step 3 of fault finding the machine.

Step 3

If our fault finding did not highlight the fault in step 2, we will continue on here. This is where you need to understand how the machine works and what it is supposed to do. Is there sequence starting of equipment i.e. is there an oil pump that needs to be started before the machine can be started. Are there any other interlocks on the machine etc.? Once you have this information, proceed to isolate all power from both the control and power (motor) circuits. PLEASE, Never Work On A Live Board!

Connect one lead from you continuity tester on the load side of the control circuit breaker and make sure you are set to the Ohm Scale. What we will do now is take the other lead and follow each wire in the circuit. Have a look at the following rough sketch.

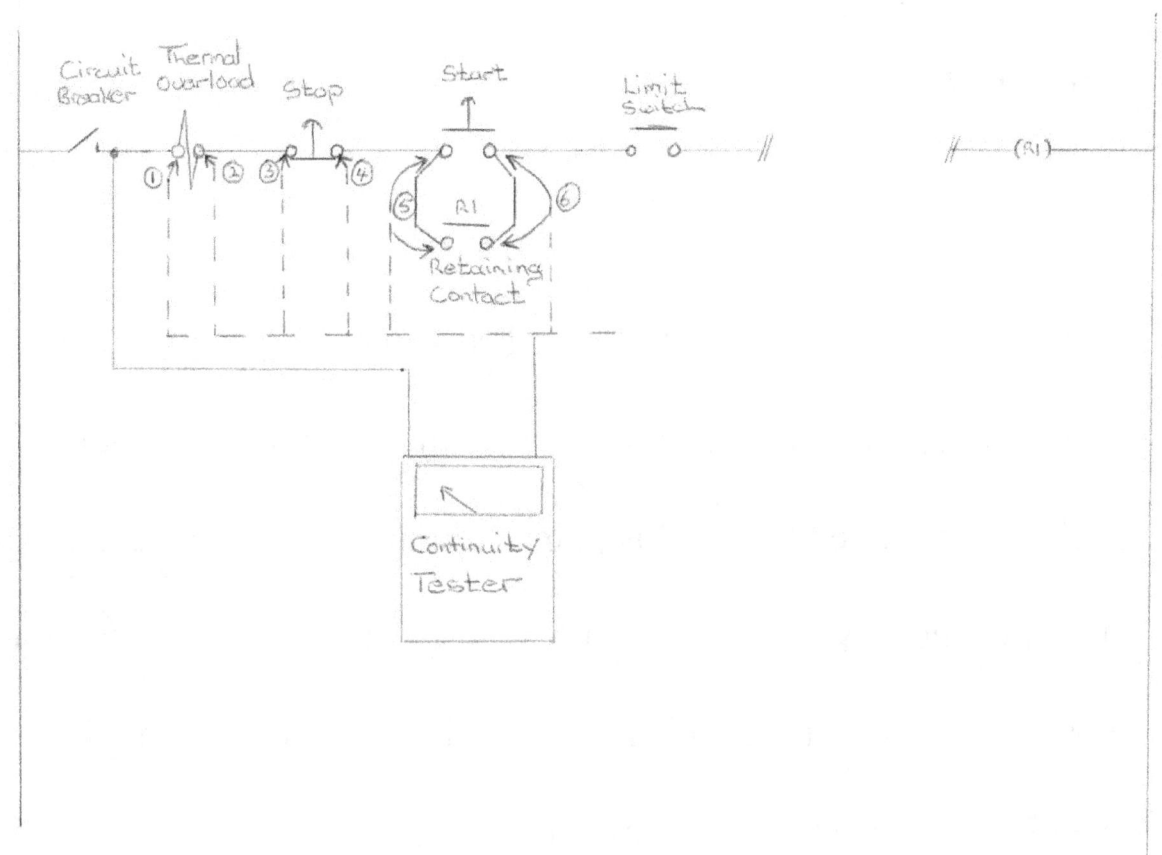

Notice the logic? You simply start at the beginning of the circuit and work your way through. With the lead not connected to the circuit, go to position one shown on the sketch. If there is no break in the wire, you will obtain a reading of virtually 0 ohm. Now we will move to the other side of the thermal overload and look for the same reading, then on to the stop button. Again, we test both sides of it. Don't be tricked into thinking you see the wire from one connection to the other and as such there simply can't be a fault. Test it and make sure. Trust your instrument!

Where the picture changes a little is when you get to position 6. When you test here, you must press the start button to obtain a reading. If the circuit has tested fine to this point, I do make one change and that is to move the "fixed" lead from my tester to position 6 and continue on from there using the same principal. I do this simply so that I don't have to keep on pushing the start button every time.

Keep on working your way through the circuit until you find the fault. Even when I found it, I tend to continue through to the end of the circuit as there may be more than one fault. Trust me, it does happen.

This method of fault finding works well even if you do not have a circuit diagram. What you need to do is physically trace the wire from one connection to the next. This is where you need to make a drawing of the circuit as you trace it for future reference, and to make it easier for the next Electrician that has to work on the machine. Seeing that you have now found the fault, let's proceed to the next step.

Step 4

- Repair the Fault.

- Test continuity after repair to make sure the fault is cleared.

- Close/replace all covers and/or guarding you may have removed.

- Inform the relevant manager you are going to restore power to test the machine.

- Ensure it is safe to do so and restore the power to all circuits.

- Get the operator to start the machine and check it is operating correctly.

Step 5

Once everyone is happy that the machine is working as it should, stop it and clean the area in which you have worked. Never leave bits of wire, insulation tape and the like behind on site.

Step 6

Inform the relevant manager that you have completed the work and the machine is ready to be put back into production.

This last section is almost as important to me as finding the fault in the first place.

Analyze the cause of the breakdown to see if there is any way that it can be prevented in future. It is not enough to merely find the fault, fix it and then walk away from the machine only to have the same fault again.

I can almost hear someone saying the limit switch failed because of the conditions it has to operate in? This might be the case. What we need to look at now is whether this is the best type of switch to use in this application or is there a better alternative? Maybe not, but I still feel the need to analyze the fault and look for ways to improve machine functionality and reliability.

Bell Tester

Here is a handy addition to your toolkit that you can build for yourself. It is a simple unit that we will call the "bell tester" You can build that and mount it into a nice enclosure and it is made as follows:

Requirements

9v Lantern battery

9v Buzzer/sounder

Enclosure

Test leads

Method:

Attach the negative from your buzzer/sounder to the negative of your battery.

Feed the test leads through the enclosure and connect one to the positive of the battery and the other to the positive side of the buzzer/sounder.

Mount the buzzer/sounder to the enclosure.

Touch your test leads together to test.

There you have it, a neat tester to assist you with testing continuity. This unit is very handy when you need both hands and you are in a noisy area. The sounder on most continuity testers are very soft which makes it awkward in a noisy environment.

Fault Finding Variable Speed Drive

Quite often you will come across an installation that has a variable speed drive in the panel. When you need to find a fault here, check the display on the drive and act on it. It is very easy to ignore the message and assume the drive is faulty. Fact is, normally the drive has just saved you a lot of time in fault finding.

Read the error code on the drive and if you are not sure what it means, check the manual. If you do not have a manual on site, the internet becomes a very useful tool as you can search the drive and get access to the manual. Just a quick tip, if you installed the Variable Speed Drive and it displays the earth fault, check your earth connection to the drive. Once you secured the earth connection, the fault will be cleared unless you have an issue downstream in which case proceed with the following:

Let's assume it indicated that you have a ground fault. This will immediately point to the motor or cable connected to it. We can determine very quickly which it is but: Proceed with Caution! Power the drive down and isolate the main switch. As usual, make sure you follow the correct lock out and tagging procedure. Once you have allowed the drive enough time to de-energize, open the terminal and disconnect the motor cable from the drive. **This is extremely important. Never use an insulation tester to test the motor with the drive still connected. This will cause serious damage to the drives electronic components and more than likely destroy the drive.**

Once you have the cable disconnected, you can perform an insulation resistance test as follows:

Connect earth lead to ground.

Make sure your tester is set to 500V.

Touch the red lead to ground to make sure your tester works.　　This should return a zero reading.

Now test each of the 3 leads of the motor cable.

If you get a reading down to earth it indicates you have a fault on one of the downstream components.

There should be an isolating switch at the motor. Turn this switch to the off position and repeat the test.

If this test indicates no fault, we have just proven the cable to be good up to the isolating switch.

Open the terminal cover on the motor and disconnect the cable.

Now test the motor as described in Chapter 2 of this manual.

95% of the time this will be where the fault is. If however the motor does test clear, test the cable between the motor and the isolating switch.

Point to Please Make Sure of:

Ensure you follow the correct testing procedure before touching/opening any switches or motors to make sure there is no power to any piece of equipment you are going to work on. You may think it is silly that I keep saying this but I know what it is like when the pressure is on. Your mind is racing ahead and already working on the fault and possible scenarios because you want production running again. The time it takes to work safe is no longer at all. In fact, that peace of mind you have would allow you to concentrate on the actual fault finding and get it done quicker.

Should everything downstream test clear, you will no doubt have a faulty drive unit and you will need to replace it. Unfortunately it does happen and the drives on the market today range from quick and easy to install through to complicated. You will need the manual for the drive to follow the setup of all the different parameters. If you have a good relation with the supplier, they may be able to assist you with this. I have had the expert on the phone with me whilst setting the parameters on a number of occasions and this saves you a lot of time. These guys are experts in their field and they know their product. I would suggest you have a look on the site you work at and note the brands of VSD's they use. Make contact with the suppliers and see what courses they have available. They often run free courses that will help you a lot and could save you a lot of headaches.

That covers in principle the earth fault. What if it was an overcurrent fault? Your first step is to do a visual inspection of the machine and in particular the section driven by this motor. It may be the fault is due to a seized bearing or something similar. It could also be caused by motor bearings that failed. Either way, if you can't find it with the visual, remove the mechanical load from the motor and run it on its own providing you can do so safely! With the motor running, listen to the bearings and take the current reading of the motor. This will quickly indicate whether the fault is on the motor or mechanical equipment connected to it. Also, check for any play on the motor shaft. This is a dead giveaway for a failed bearing.

Chapter 2

Motor Testing and Connections

As mentioned before, when we do a motor test, it is important to include the mechanical side as well. Let's look at performing a simple motor test. For this test I will work on the basis that the motor has been removed from the installation and is sitting on your workbench.

1. Take the motor shaft and turn it to make sure it spins free. If it does, listen to the sound of the bearings. Be on the lookout for a grinding noise which will indicate failed bearings. As a rule of thumb for, if the motor has been removed from service, change the bearings anyway, providing it passes the electrical test.

2. Check for any cracks on the mounting feet, flange (if flange mounted), end-shields terminal box etc. This sort of damage will

lead to failure when you least need it and should be repaired before putting the motor back in service.

3. Check the condition of the fan and fan cover. Always ensure the fan cover is in a good state and covers the fan completely. This is an important safety aspect as you don't want anyone getting a finger in there when the motor is running.

4. Only now will you move to the Electrical part of the test which is done as follows:

4.1 Connect one lead of your insulation tester to ground. Ensure you have a good connection by pressing the test button and touching the other lead onto a different spot of the motor casing. Remember to set your meter to the 500V scale. If your connection is good, you will get a deflection on your instrument the same as you would by touching the two leads together, i.e. short circuit.

4.2 Leave the one test lead connected to earth and attach the other test lead on each end of the different motor windings one at a time. Write down the reading obtained. Although Wiring Regulations state that the reading needs to be higher that 1Mega Ohm, I prefer the reading to be above 2 Mega Ohm and best case scenario, a reading of infinity. Now also test between windings for a short. Connect one test lead to a winding and the other to another winding. Again I will be looking for a reading above two Mega Ohm and still prefer the infinity reading. Proceed to do this between all windings. If the readings obtained indicates that there is no short down to earth or between windings, proceed to the next step.

4.3 Change the scale on your insulation tester to 200 Ohm, disconnect the test lead from the earth position and now we will test the continuity of each winding. This is done by testing between the two ends of the winding and recording the reading. Remember, you are now reading on the Ohm Scale! Once you have recorded the readings of all three windings (this test is based on a three phase motor), you need to compare the readings obtained. Look for the readings to be at least within 10% of each other. If it is within this range, you can deem the motor Electrically Sound and ready to be put back in service providing all mechanical areas are also repaired and bearings replaced.

Just as a short explanation, each winding has a starting and ending point. These are numbered as 1 and 2. You will see on the winding or the terminal block that they are marked as U1, V1, W1 and the other side U2, V2 and W2. Make sure you keep all the ones and two's in the right spot! You don't want to get one of them the wrong way round when connecting the motor to power. So what do you do if you get handed a motor with six wires dangling loose in the terminal box and they are not marked? Don't despair, I will show you a very easy way to test them and get the numbers sorted in the next section.

Establishing the start and end of motor windings

As mentioned in the previous post, you may end up with a motor that has 6 leads dangling from the terminal box and they are no longer numbered. Your task is to test the motor and prepare it to be put back in production. This may seem a bit daunting but it is actually quite easy to do.

Before you start with anything on the numbers, perform the motor test as per the previous post. To do this test for the Electrical side, just group your windings in pairs and don't worry about which is one and which is two. Just find the pairs by testing the continuity of the windings with your Insulation Tester and remember to use it on the Ohm Scale. Once you have the pairs grouped, you can complete the Motor Test and move on to the next step of identifying the start and end of each winding.

You will need a 12V car battery (my preference) and an analogue multi-meter for this step.

Connect a red wire to the positive of the battery and a black wire to the negative. Make sure you don't touch the ends together! Now swap the leads on your meter the wrong way round i.e. Red into the black socket, and black into the red socket. If your meter leads are permanently connected, just mark the ends with some red and black tape. (red tape onto the black lead, and black tape onto the red lead)

What we will do now is to state that we call the red leads number one and the black leads number two. Connect the other end of the meter leads to any one of the winding pairs and make sure you have your meter set to measure DC Voltage.

Now for the fun part. Hold the black lead on one end of another winding pair and touch and remove the red quickly on the other end. This will induce a voltage into the other windings and you will note a very quick deflection on your meter. If the deflection is positive, you have the numbers correct. If the meter deflects to negative, the numbers should be the other way round. Mark the motor leads at the battery pair one and two and do the same for the winding just tested.

Using the same winding for the battery, perform the test on the remaining winding.

Well done! You have just paired the windings and established the beginning and end of each using some very inexpensive gear.

Just remember to always swap the meter leads over when doing this as the induced voltage is in the reverse direction.

Typical Motor Connections

Motor connected in STAR condition

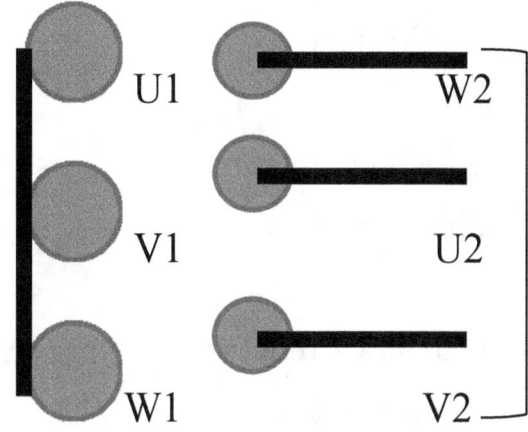

to contactor

Motor connected in DELTA condition

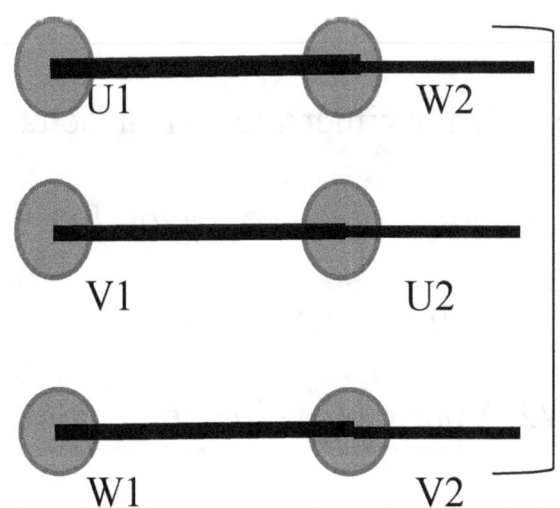

to contactor

Motor connected to STAR/DELTA starter

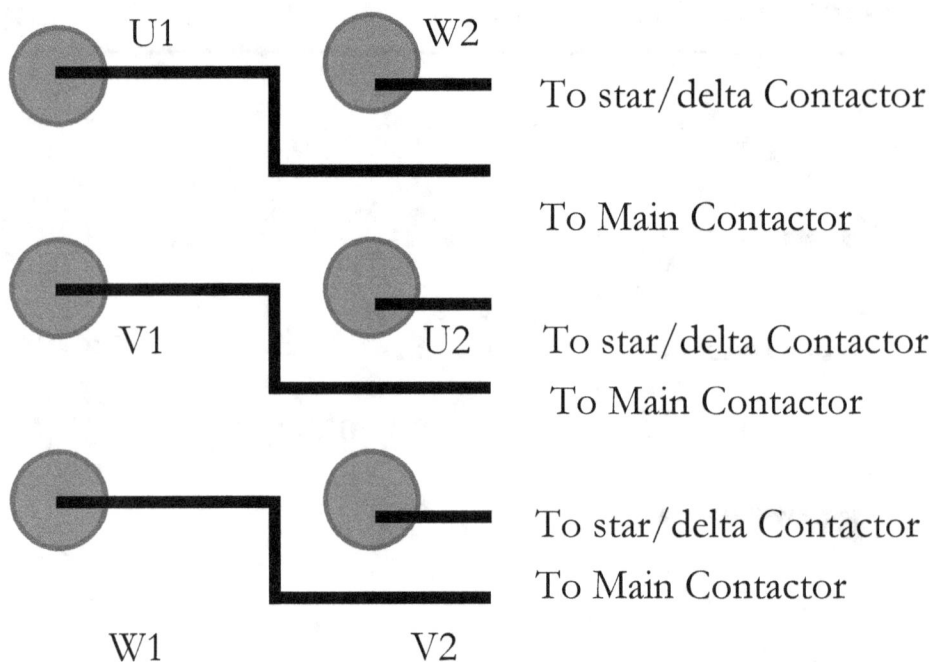

U1

W2

To star/delta Contactor

To Main Contactor

V1

U2   To star/delta Contactor

To Main Contactor

To star/delta Contactor

To Main Contactor

W1

V2

Changing the direction of rotation

If you need to change the motor direction for either the star or delta connections, change any two of the leads going to the contactor. Let's say you connected the red phase to W2 and the white phase to U2, change it to red on U2 and white on W2. Your motor direction is now changed. The only thing to be very mindful of here is the star/delta connection. **If you have to change the direction here, do it at the bottom of the circuit breaker and not at the motor!**

The reason here is that you need to maintain the same phase to the windings as per the diagram. When the motor is running in delta, you need to have the same phase connected to U1 and W2, the next phase to V1 and U2 and then the third phase to W1 and V2!

This rule also applies when you connect your motor to a variable frequency drive. If you want to change the rotation and you do not want to do it in the drive parameters, change two phases over on the supply side to the drive and not the load side. These drives are smart and even if you change two phases on the load side, you will end up with the motor going in the same direction.

Another point of importance; Read the manufacturer's nameplate before you connect the motor. This will tell you what condition the motor needs to be connected in for a specific voltage.

You may find the following:

Condition: λ/Δ

Voltage: 415/230

This means if you are connecting it to a three phase 415V supply, you need to connect it in the STAR condition. If however you are connecting it to a 230V three phase supply, you connect it in the DELTA condition.

Be careful when you use a variable frequency drive! If the drive is a single phase input to a three phase output, it means that you will have 230V three phase to your motor and you will need to connect the motor in the delta condition.

Single Phase Capacitor Start Motor

Basic Diagram

This is a basic and easy sketch to remember and will serve you well. Draw this even before you start testing the windings and just add the values to it. Remember, the highest resistance reading of the two windings will be the start winding. When using this basic sketch, you can easily determine the connections on the motor. Big question is always: How do I change the direction of a single phase motor?

This can be done by swapping either the start or the run winding. Not Both!

If you have a look at the next sketch, you will note I have changed the start winding. For some reason this has always been my preference.

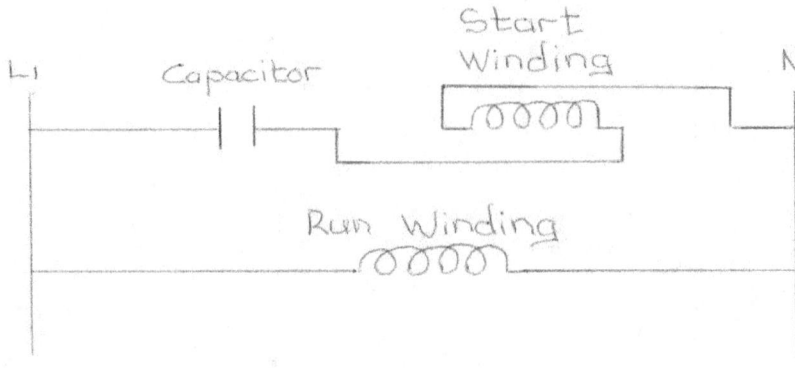

Again, this sketch is a basic thing to remember and anytime you need to work on a single phase motor, make the sketch, take the readings and you are set to go.

Chapter 3

Preventative Maintenance

It's not just Fault Finding!

What Exactly Is My Role?

For this manual, I have focused on a generic type role description for the position of Maintenance Electrician.

Take Ownership of What you do!

Most companies will tell you that your role is to ensure equipment functions to the specifications and to conduct fault finding in the case of a breakdown. In short, they are saying they don't want breakdowns but should that happen, you need to get it fixed in the shortest possible time. The real issue here is that they expect you to know what to do in order to maintain maximum up-time of equipment. This is easy enough if you have been working in the maintenance environment for some years so as to gain experience, but what do you do when this is all new to you?

Let's take a look at the requirements first.

You need to perform Preventative Maintenance to prevent downtime.

You need to perform reactive maintenance to repair breakdowns.

You must maintain Safety Circuitry to ensure machines are safe to operate.

You need to report back on equipment status, cost of repairs and efficiency.

You need to perform alterations to improve machine efficiency.

In a nutshell, that will be your primary task. The only thing you have not been shown is the How? So let's take a closer look at the things that would show you what needs to be done on a typical plant.

Prevent motor failure due to bearing failure.

Prevent equipment shutdown due to faulty switchgear.

Ensure all connections on equipment are tight.

Prevent water from entering electrical equipment in wet environment.

Prevent nuisance tripping by safety devices.

Sound easy enough right? Well this is just a basic description on the things that would form part of your task and the next step is to start going into the detail of how to do it and also the things that would make you stand out from the rest of the Electricians that stick to doing exactly what the foreman/supervisor tells them to.

As you will see, a lot of companies will employ a Maintenance Planner. The role of the planner is to schedule tasks for the maintenance crew in terms of preventative maintenance to ensure breakdowns are kept to a minimum. What you need to remember here is that the success of the Maintenance Planner is determined by the efficiency of the plant. That is the best way to measure his/her success but is it the best method in terms of being cost effective. Think about this, if you are the planner and you are going to be measured in this way, what would you do? Pull schedules forward so as to make sure you don't have downtime? That is what it is all about right? In a sense yes, but what you need to be mindful of is that this could lead to equipment being replace way ahead of the expected lifecycle which leads to the maintenance not being done cost effective. I am in no way saying this is what every maintenance planner will do, but it is a risk. So how does this affect your position? Have a look at the next sections and decide for yourself.

The Administration Side

We will start with the most important task in the role of the Maintenance Electrician. You must know what you have to maintain. Even though everything may be listed in the maintenance software, my recommendation is to make sure you do the following:

Set up a Journal for every machine you are responsible for.

Break the equipment down into sections to show every part on the machine.

Record circuit breaker sizes and, if possible, date of installation.

Record detail of contactors and date of installation.

Record detail of overload protection devices, setting of device, maximum rating of motor and date of installation, if known.

Set up a Journal for every switchboard/control board. Record every component in the board as with the units above.

Up to this point we have not done anything out of the ordinary. We have merely created a record of the equipment you are responsible for. With this information, we can now move to the next step where we start using this information in a sensible way.

Make the Journal part of your Toolkit!

Journal Example

To better explain what I mean, have a look at a sample page of a typical Journal for this action. You can alter the configuration to what style you prefer, add or remove columns and possibly even include a photo of the equipment. This is your document and you need to feel comfortable using it.

This page on the example deals with only one component. You have to include a page for each individual component associated with the equipment.

In other words, you will have a page for the contactor, another for the overload and so you complete the full list. These lists go into the folder and are kept together.

Include the circuit diagram for this installation as well, and while you are at it, leave a copy in a folder inside the cabinet!

Waste Water Pump 1

Power Supply From Distribution Board X1, Circuit Breaker 13, 15, 17   63A 3 Phase + Neutral

Cable size 25mm² 4Core + Earth

Motor

Size	Rating	Speed	Bearings Drive End	Bearing Non Drive	Mounting	Date Installed
15Kw	34A	1440	63092Z	63092Z	Foot	Unknown

Service History

Date	Work Performed	Technician	Signature

What is Preventative Maintenance?

To put it all into one sentence, your job is to carry out preventative maintenance to all plant and equipment in a cost effective and efficient manner so as to ensure downtime is kept to a minimum. We have now had a look at the tools you can create to streamline your planning side of the work, so we will now take a look at some of the tasks to be performed.

Take a look at the example we used for the Waste Water Pump 1. Start at the Supply Board where the circuit originates from.

Here we have a 63A 3 pole Circuit Breaker.

- Perform correct Lock Out Tag Out procedure

- Move to the Control Board and perform test to ensure all power to this board is OFF

- Check all circuits; make sure you do not have multiple supplies to this board. Only once you have ensured all power to this Control Board has been isolated, can you proceed.

Now you can start with the Preventative Maintenance.

1   Remove the cover from the motor terminal box.

2   Test Motor connections again before touching.

3   Check that all connections are clean and secure.

4   Perform complete motor test and record all readings.

5   Check cable glands to ensure it is secure.

6   If you are satisfied with test results, replace cover and make sure you maintain the seal.

7   Check the cable from the motor to the local disconnecting switch. Make sure there is no damage to the cable, it is securely fastened and protected from any possible mechanical damage.

Record findings and any remedial work done.

8   Remove the cover on the local disconnecting switch and test all conductors for power.

9   Inspect the enclosure and cover for any visible damage like cracks, damaged seal etc. Record findings and any remedial work done.

10   Inspect the switch for any visible damage, clean and tighten connections. Record findings and remedial work done.

11   Check the continuity of the earth conductor. Record reading.

12   Once complete, close the switch and again, make sure you maintain the seal.

13   Check the cable between the switch and the control board for any damage, ensure it is securely fastened and that all cable glands are secure.

14   Open the control board and test all active conductors. Be on the lookout for multiple supplies in the board.

     Make sure to also test the control circuit. Never trust the fact that you have turned the supply off, locked it and placed your tag on the lock! Always test before you touch!

15   Once you have established that there is no power on the equipment, start by performing a visual inspection. Be mindful

of discoloration of leads at the terminations as this is a giveaway for a loose connection and/or faulty switch. Record your findings and any remedial work done.

16   Now start on the power circuit and clean and tighten all connections.

17   Check the circuit breaker ratings against the cable size to ensure all conductors are protected against overcurrent. Pay particular attention to control circuit cabling tapped into the load side of the main switch as I have come across this on a number of occasions. Should you find this, carryout the necessary remedial work. Record all findings and work done.

18   Check the setting of the overload protection device and record the setting. Make sure it matches the maximum rated current of the motor.

19   Move to the control circuit when done and go through each device connected in the circuit.

Check for any damage, wear, connections and carry out necessary remedial work. Record all findings and work done.

20   If your company performs regular injection testing, record the date on which it was last done.

If not, consider the implementation of such a schedule or at least perform a thermal imaging test on the equipment when under load.

21   Once you have completed all the tests, inspection and remedial work, perform a complete installation test and record all readings. Where it is a Legislative requirement, complete the Electrical Certificate of Safety as stipulated.

22   Notify the relevant manager that you have completed your work and will be restoring power.

23   Once the unit is running at full load, record the current drawn by the unit, check this against the overload setting and the motor rating.

24   With the motor running, it is also the perfect opportunity to listen for noisy bearings should you have missed it when performing the motor test.

You will quite often hear the Electrician saying that the bearings are not his responsibility as it is mechanical. I would rather make that part of my work rather than having to replace the motor because of bearing failure!

Up to this point we have not done anything out of the ordinary and could easily sign off on this job as complete. As we are aiming to go the extra step, here is what we do next.

- Calculate the number of starts/day of the motor.

- Check the switching duty of the contactor. This should be available from your supplier.

- Make sure the contactor duty cycle is correct for this application and if not, replace it with the correct unit.

- Should the duty cycle be correct, find the life cycle for this contactor. (number of switching actions)

- If you have the date on which it was installed, calculate the time left before recommended replacement and schedule the task.

Without the date it is a bit harder as you may have to estimate a time. This is where Thermal Imaging will be of great benefit. By doing this you unfortunately do run the risk of it failing before the scheduled date. What is important is to make sure you record everything you do.

Although we have looked at contactors in the above, the important thing here is the method and record keeping. You can apply the same method to establishing the expected life cycle of switching relays.

Try to avoid the run to destruct method as this is the most costly form of repair.

Make Use of Pictures

When you set up your Journal, don't discard the use of pictures/photos. This is a valuable item to add to your Journal. It is always useful to have a visual of the equipment to refer to. It goes a long way with any audits your company may have etc.

**Control
Board D5**   *Enter a description of the board, number of
devices etc. for use as a quick reference.*

Summary

We have now worked through a nice and simple procedure on a basic installation. The most important thing here is to set up your Journal for each individual piece of the installation and machines. The procedure remains the same.

Once you have done this Journal, make sure you use it and update it every time you perform any work. The one thing we have not discussed here is the analysis of each failure.

Never accept everything as normal. Always try to establish the cause and then find ways to improve it.

- When you are looking at the duty cycle of the contactor/s, keep in mind the environment in which they are installed. Things like ambient temperature, cabinet temperature etc. All these items have an impact on your contactor performance and/or life cycle.

- You will also note that most insurance companies now insist on thermal imaging being done on all electrical components as they regard the electrical installation to be the highest risk factor. In doing the thermal imaging survey, you need to compile a full report on this survey.

- This is a document that you need to prepare in full and keep it on record for future reference. When the assessor does a site inspection, he/she may ask for proof of the thermal imaging inspections. This is a stand-alone document and does not go into your journal. All you do in the journal is to record the date on which it was done, the findings and any remedial work that was performed based on the findings. Importantly, you must also show what actions were taken in your report. It will make no sense if you established that there was a problem but did not rectify it.

- If your company does not have a Thermal Imaging Camera, there are a number of specialist companies that can do the survey for you and they will also produce a full report on the site.

- The main point of difference here is that you will need to work through their report to rectify any issues after you received it as they do not necessarily conduct these repairs. Again, when you perform this work, make sure it is recorded. You will also find that a lot of insurance companies prefer the third party survey over that of the in-house report.

- There are also a number of other items that need to be done on a periodical inspection routine. These items are part of the employer's duty to provide a safe workplace and it is a legislative requirement. These include:

Testing of all Earth Leakage Protection Devices. (also referred to as residual current devices) It is important to perform the test as prescribed by your relevant Act, and to record all test results. It is also a requirement to keep a log of all these devices and test results.

Testing of Emergency Lighting. Again, you must keep a log of this.

Testing of all Emergency Exit Lights. As above, keep a log on all of these tests.

Testing and Tagging of all portable appliances, extension cords etc. Again, you need to maintain a log of these units and tests performed.

If you look at your particular site at which you work, how much time would you be spending doing all these tests? If you add all of these to your workload, do you feel you can do this in-house and maintain all the required documentation as required by Legislation?

I ask this question not to indicate you should not do it, but rather to consider using a specialist company that will maintain all of these inspections/tests for you, as well as provide you with a comprehensive report on completion.

Another service they would offer is to notify you in advance of items that need to be tested again. To me, this makes a lot of sense and you may be surprised to see that it is cost effective to have an external provider perform these tests/inspections as opposed to doing them in-house.

Chapter 4

Safety of Machinery

With the changes in Legislation, this has become a crucial part of any business. The employer has to ensure that all machines conform to the latest Standards on Machine Safety. The safety of the machine is not isolated to the operator only. We need to consider the operator, anyone in the vicinity of the machine, maintenance staff as well as the cleaning staff.

I would advise that you obtain a copy of the Standard and read it, use it as a reference manual and always keep it updated. The importance of this is huge. We do not want any person getting injured in his/her workplace and as such, we must do everything humanly possible to safeguard every piece of equipment so as to minimize the risk of injury.

How do you know what to do?

We need to start right at the beginning and consider the machine as it stands. This is where you would need to have the Risk Assessment done on the machine. Every aspect is looked at and things to be considered are the actions of the operator. Always keep asking What If? As mentioned before, there are specialists in the field that can do these assessments for you and they will generally make recommendations as to how you can improve the safety of the machine.

The machine will be assessed and a Risk Category will be defined. This category is based on the severity, frequency and likelihood of an incident. This is also the category that will determine what type of safety system you need to install on this machine.

With the changes over the years we are now heading to the point where we need to also build into the assessment the reliability of the safety equipment. In other words, apart from ensuring the safety of the machine, we also want a system that aids production by not having constant failures.

With the category assigned to a machine, the level of protection will vary. This means you will range from a very basic pull cord system to a fully monitored system. This is typically where you will have some small stand-alone systems on a few machines or a safety PLC that sits over the top of all the field devices.

Important point to remember here is that you are not allowed to run your safety circuit on the same PLC that runs the machine. It has to be an approved and dedicated Safety PLC.

The switches and/or sensors used on a safety circuit are also specifically designed for that. Do not use a normal limit switch to act as a safety device. Also stay clear of any normal stop buttons being used as emergency stop units. The Emergency Stop button also needs to meet the specific Safety Standard.

It all sounds a bit daunting at first, but once you have the system set up on your machine, you will see the benefits of it. **One important point to remember is that every time you add a safety device; assess the operation of the machine again, as by adding it, you may create a new risk. This is extremely important.**

Also, don't just focus on the operator; you must design for operator, maintenance and cleaners. It is a common mistake to just focus on the operator. The problem arises then when you need to perform maintenance to the machine. How does the maintenance staff access the machine, how do they test it and what impact will the operation of a safety device have on them?

The same goes for the cleaning of a machine. These guys need access to get to the machine and again the safety system will have an impact on them so please make sure you consider their safety as well.

How it Works

You now have a machine that was assessed as a category 3 machine in terms of risk. This means you will have dual circuit safety switches on the machine and all of these safety switches are wired back to a Safety Relay

Once the machine is running, it is necessary to check each of these safety devices/switches to make sure it does what it is supposed to namely; shut the machine down.

When the machine is powered up, the Safety relay checks the change of state on each of the switches. This is how the monitoring system works. If there is no change of state, it deems the switch faulty and indicates the fault. If you look at the front of most safety relays, it has a green LED for each input to show the circuit is healthy or a red LED in the case of a fault. When this LED is red, the machine will not start.

These relays can also wired with a reset button. In other words the gate was not shut properly and the machine failed. You now close the gate but the machine will still not start until you have pushed the reset button.

The safety switch will have two contacts and both will be wired to the relay. The relay checks the state of these contacts and then provides two outputs to two safety contactors; normally they will show them as K1 and K2 on the manufacturer drawing.

When you wire your control circuit, you will wire the normally open of K1 and the normally open of K2 in series with your circuit. That means that both of these contacts are made before you are able to start your machine. The question is why do we need two? This is done in the likelihood that the contact used "welds" in the closed position. The likelihood of both doing that is very slim. Should it happen to one of the contacts, you should find that during your routine inspections which we will discuss next.

You will note that I am not providing you with a wiring diagram for the safety relay. This is due to the number of safety relays available and also the number of different configurations pending the category and safety switches being used. These diagrams are available from the manufacturer and you simply select the drawing that will suit your application.

Of utmost importance is that you make sure you wire the safety circuit to the category the assessment came to or higher. There is nothing wrong by installing more than the standard calls for, but if it was possible to improve on the safety, it was most likely missed during the Risk Assessment.

The picture shows the reset we spoke of earlier. Look at how it happened:

- Door opened;

- Safety Switch opens and machine stopped;

- Door was closed but machine will not start;

- Safety relay checked state and provides output to reset indication;

- Once reset, machine can be started.

You will note the Emergency Stop is missing the yellow faceplate in this picture which means it does not meet the Standard.

What You Need To Do

Although we are adding all these safety devices to ensure the safety of the operator, maintenance staff and cleaning staff, we need to maintain these devices in a safe working order.

As we have done earlier in setting up a Maintenance Journal, you will need to do the same for every machine that has these safety devices fitted to it. The first thing you need is a Standard Operating Procedure to conduct the tests on these devices. This needs to be an accurate description of how to perform the test, and needs to describe the procedure for each of the safety devices on the machine.

• You must record every test you do, the result of the test, distance of travel of the safety switching before activation of the safety and also any remedial work performed. So what do I mean by "distance of travel"? It simply means that in the case of a door for instance, how far I can open the door before the safety circuit drops out and the machine stops.

What you need to look at here is whether the machine will be completely stopped by the time I could reach any moving parts of it. This does not mean you need to try it; there are formulae to determine the distance between the guard and the machine, the speed of the movement and the time for the machine to stop.

Once again, your Risk Assessment Specialist will be able to give guidance here in terms of the relevant standard.

Performing the Inspection on the Machine:

• Inform the relevant manager that you will be conducting tests on the machine.

• Stop the machine and isolate only the power circuit. For the first tests, the **control circuit** needs to be on.

• With the machine stopped, open a guard and check whether the fault has registered on the control circuit Safety Relay.

• If the fault has registered, close the guard and reset the safety relay then move to the next safety device and repeat the operation.

• Continue in this way until all safety devices have been tested and you have established that they are all in working order.

• Once satisfied that all the devices are functional, start the machine and then repeat the tests by very carefully opening the guard with the machine running. This time, move the guard extremely slow as you want to measure the distance it moved before the safety switch stopped the machine.

- Again, repeat this until every safety device has been tested and travel recorded. Note that you do not record distance of travel for any Emergency Stop unit.

- Ensure your findings are in line with the standard. If they are, you can sign off the work and inform the relevant manager the machine is safe for production

The above tests are typical of what you would do on a weekly basis. As for the Monthly test, include only one thing before you proceed with the above steps

- Isolate lock and tag out machine at the point of supply

- Test all conductors to ensure machine is safely isolated

- Check each cable to safety devices for any damage

- Clean and tighten all connections on safety circuit.

- Test continuity through all contacts of every safety device to ensure they are all functioning.

- Once you have recorded all findings and work done, return to the weekly procedure and complete the inspection.

Please Note: This is how I perform these inspections and in no way am I stating this is the only way to do them. The point here is that you need to ensure that the safety switches work, and that you record every task performed on this machine.

Should there ever be an incident on a machine, you will need to prove how you inspected this machine and these documents are crucial. The most important thing for you to be aware of is that you will be forcing the machine into a fault whilst running.

You must only do so after a full inspection of the safety devices to make sure they work. Make sure that it will be safe for you to do this!

In other words, is there sufficient distance between the moving part and the guarding that you plan to move? If it is not possible to do this safely, you need to re-assess the safety devices as this means they are too close to the moving parts in terms of the standard?

Once you have completed all the tests and deem the Safety Circuit to function as intended, you can proceed to put the machine back into production. As with procedure, notify the relevant manager that you have completed your work and will be restoring power to the machine.

What if it fails?

Should any part of the safety circuit not function as intended, you never put the machine back into production unless the fault has been rectified and tested! **Never bypass a safety device!**

If you establish that there is a fault, proceed to fill out a Defective Machine Report. This will serve to inform the relevant Manager that the machine is deemed unsafe and may not be used. Important thing here is to do this report even if you are going to remedy the fault immediately. We are again proving that your procedures are being followed and we maintain our records to prove this!

Make sure that you have performed the Lock Out, Tag Out procedure and the machine remains in this lock out state until the fault has been repaired and tested

Test Sheet Sample

Cover Page

Machine Name:

Location:

Standard Test Procedure

• Inform the relevant manager that you will be conducting tests on the machine.

• Stop the machine and isolate only the power circuit. For the first tests you need the control circuit to be on.

• With the machine stopped, open a guard and check whether the fault has registered on the control circuit Safety Relay.

• If the fault has registered, close the guard and reset the safety relay then move to the next safety device and repeat the operation.

• Continue in this way until all safety devices have been tested and you have established that they are all in working order.

• Once satisfied that all the devices are functional, start the machine and then repeat the tests by very carefully opening the guard with the machine running. This time, move the guard extremely slow as you want to measure the distance it moved before the safety switch stopped the machine. Record the distance moved before activation of Safety Device as well as the time before the machine came to a full stop.

• Again, repeat this until every safety device has been tested and travel recorded. Note that you do not record distance of travel for any Emergency Stop unit, only the time for the machine to come to a full stop.

• Ensure your findings are in line with the standard. If they are, you can sign off the work and inform the relevant manager the machine is safe for production.

Test Record

Machine:

Location:

Date Tested:

Technician Name:

Technician Signature:

Device	Pass	Fail	Time to Full Stop	Distance Travelled Before Stopped	Remarks
Emergency Stop 1				NA	
Emergency Stop 2				NA	
Safety Limit Gate					
Safety Limit Guard					
Safety Mat				NA	

Please note that your relevant Standard may require more information and/or a different format. Approach your Local Authority for advice as you will find them quite willing to assist as you are ensuring Machine Safety Standards are correctly implemented. You can set up a test sheet for a machine and show it to the person in your area for recommendations to improving it.

As with everything, we are now aiming to improve not only the Safety, but also the way in which we do things. So, can we **do more** than simply follow the procedure as explained and sign off the test?

The answer is YES!

Consider adding the following items to your checklist:

Does the Emergency stop meet the relevant Standard?

You may have had the correct stop in place but at some stage it may have been replaced. The unit in stock was not an approved emergency stop but it was installed to get production going!

Is the Safety Limit on the gate an approved safety device?

Record the part number of the unit installed. If it has a serial number, record that as well

Is this the best switch for the application or can it be bypassed?

Chapter 5

Circuit Design Methods

This section will touch very briefly on the topic of circuit design. The idea is to give you a starting point from which you can further improve your skills.

As with any machine, the control of it needs to be defined before you can start designing a circuit. This is once again where we need to make sure we do not try and follow shortcuts as in the end, they will make you lose time.

Before you start with the circuit diagram, you need to have a full understanding of what the machine needs to do. Start by writing a description of the entire system.

As an example, we will look at a design where we have four conveyors that need to be semi-automated. So let's start with the description of the system.

1 Machine components

Conveyor 1

Conveyor 2

Conveyor 3

Conveyor 4

2 Component Function

Conveyor 1

An empty container is loaded onto this conveyor and the operator pushes the "LOAD BUTTON" for the conveyor to move the container to the end of the conveyor. Once the container gets to the end, it is detected by a limit switch and the conveyor stops. This conveyor will be named: LOAD CONVEYOR.

Conveyor 2

If there is no container on this conveyor, it will detect the container on the Load Conveyor and start automatically. Once this conveyor is running, a signal will be provided to the Load Conveyor to start and transfer the container to conveyor two. Once the container has been transferred, the Load Conveyor will detect that it has no longer got a container on it and stop.

With the container now moving forward on conveyor two, it will move the container to the end of the conveyor where it will be detected and the conveyor stopped. This conveyor will be named: ACCUMULATION CONVEYOR.

Conveyor 3

The same will apply as with the previous two conveyors. If there is a container present, the system will wait until it is moved. Once there is no container detected, the Accumulation Conveyor as well as conveyor three will start. The container will be transferred to conveyor three which will stop the Accumulation Conveyor. Conveyor three will run until the container is detected at the end of the conveyor where it will be stopped.

Once it is in position, the container is filled by an operator. When the container is full, the operator will push a button to move the container to the last section. This button will provide a signal to conveyor four as well as conveyor three to start. This may only happen if there is no container present on conveyor four. This conveyor will be named: FILL CONVEYOR.

Conveyor 4

This conveyor will receive it's start instruction from the operator pushbutton. Providing there is no container present, it will start at the same time as the Fill Conveyor and move the container to the end of conveyor four where it will be detected and the conveyor stopped. Once the container has moved into position on this conveyor, it needs to provide and audible Alarm for a time of 10 seconds to warn the unloading operator. This conveyor will be named: UNLOAD CONVEYOR.

Once you have this basic description, proceed to draw a sketch of how the systems will fit together. On this sketch you can then start adding the switches you may need to perform the relevant tasks.

Notice how we have not drawn any part of the circuit yet, but already we have created a typical instruction set in the description?

Diagram

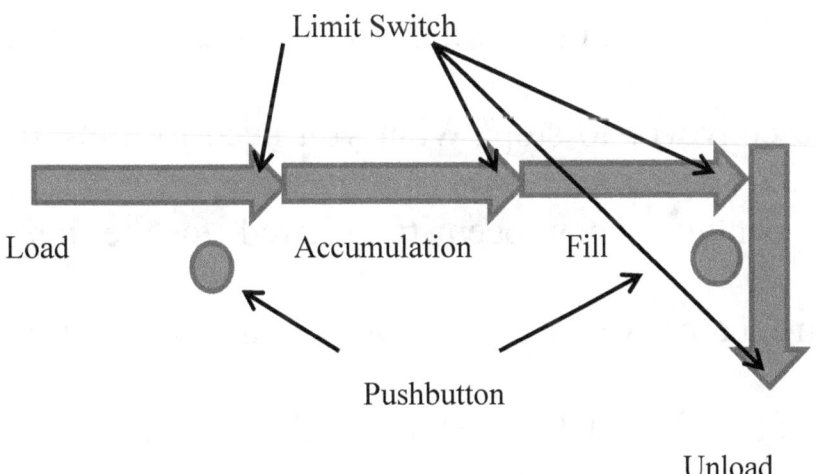

From this diagram, you can see what hardware you will need for the circuit. There is not a lot in terms of the field devices, 4 limit switches and two pushbuttons. The other items you need to consider here are the four motors and that means four disconnecting devices or as it is commonly referred to, isolating switches. Now read on to see what we have not shown and you will understand why we do this exercise before we draw any circuit!

Now try to picture the movement of the container on the line. It is sitting on the limit and is ready to move forward. The conveyor starts and it moves away, thus the limit switch will reset. This will immediately cause the conveyor to stop. What you need to think of here is whether the container has been transferred to the next conveyor before the in-feed conveyor stops. This will also depend on the exact location of your limit switch, the shape of the container etc.

As you can see, there are a lot of things to consider before stringing together the control circuit and then hoping that it will work. This machine may need to be equipped with variable frequency drives.

This will allow you to adjust the acceleration and deceleration times. This will allow you sufficient time to transfer the container before the conveyor is fully stopped.

What needs to be looked at as well is the machine safety. There will be a need for Emergency Stop switches, a stop and start station for the entire system. Maybe you want to build a diagnostic feature into the machine. In other words, maybe some indication lamps on every limit switch.

Notice how many extra items we have just added. Now imagine that you have already done your circuit design and started wiring it only to now realize you have to change it!

Only now are you ready to start working on your control circuit. You have established that you need the safety circuit and that will be wired directly to the Safety Relay. You will then use the auxiliary contacts on the two safety contactors in your control circuit.

There will be a central stop and start station to start the system at the start of production. As this machine is a semi-automated system, you have decided to provide a visual indicator that the system is "ON"

As the system start will not be turning on any conveyors or other equipment, it means you will use a relay or contactor to be activated and then provide power to the rest of the circuit.

Take a bit of time and work through the circuit on your own and see if it will match the description we have drawn up. Does it work, what is missing, how can it be improved? This is all part of designing a circuit and making sure it will work before you rush off and start wiring it.

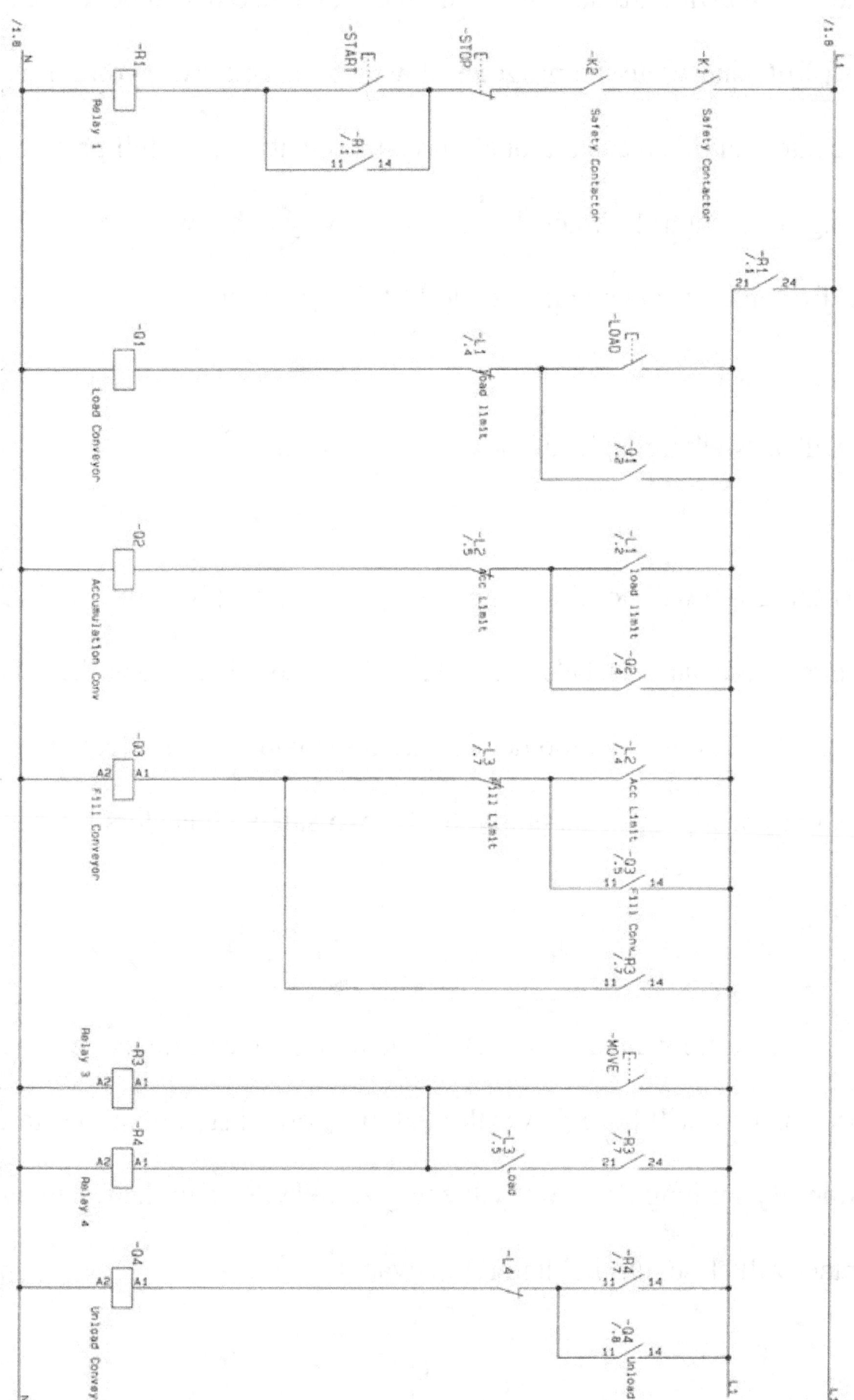

81

When tasked with this sort of circuit, the preference would be to use a small PLC and write the program, download it and away you go. In the context of this manual though, it is important to get a full grasp of setting up a control circuit the traditional way. This will eventually pay dividends when you move on to PLC Programming.

Now do a "walk through" of the circuit as drawn.

First thing you will notice is that coming off the supply at the top, we go into two contacts labeled as Safety Contactor. The reason here is that all the Emergency Stop devices are to be wired via a safety relay. This relay will provide an output to the two safety contactors K1 and K2.

Assume the safety circuit is on. Both the normal open contacts will be closed and we will have power through the normal stop onto the start button. By pushing Start, we will energize relay R1 and the retaining contact will close thus holding R1 activated.

With R1 energized, the normal open of R1 will close, and as such provide power to the line for the machine side. If you look at this line, you will note no contactor can be energized as the entire line has normal open contacts.

Now the operator pushes the LOAD button. Power goes through the normal closed contact of Limit 1 and Q1 is energized. As before, the normal open contact of Q1 retains the circuit and your conveyor is now running thus moving the container forward.

Once the container reaches the limit switch, the contacts will change state. That means the conveyor is stopped as the normal closed contact has opened, but at the same time the normal open contact closes on the next section of the circuit.

This now creates a path through the normal closed of Limit Switch 2, and energizes Q2, which as before, has a retaining contact and the second conveyor is now running and moving the container forward.

Once the container gets to the end of the second conveyor, it activates Limit Switch 2 and the conveyor is stopped with the normal closed contact now opened.

Again, this closes the normal open contact on the Fill Conveyor section, and Q3 is now energized. This conveyor now runs and the container moves to the end of the line where it activates Limit Switch 3. Now you will notice that apart from the retaining contact in parallel with the normal open of L2, we also have a normal open of relay R3 drawn in to bypass both the limit switches to this coil. The reason here is that we do not want this conveyor to start the unload section until the operator pushes the MOVE button.

So this means the container has stopped on the limit and the conveyor was de-energized. The container is now filled, and once the operator is happy, he/she pushes the MOVE button. This energizes relay R 3, which will start the Fill conveyor as well as the unload Conveyor. This happens with the normal open contact of R 3 closing and energizing Q 3, relay R 4 has also been energized which in turn energizes Q 4.

What you need to observe here is that the retaining contact on both R 3 and R 4 is done different. There is a normal open contact of Limit 3 in series with the retaining contact! This limit has changed state when the container activated it, and will again change state as the container is transferred to the unload conveyor thus stopping the Fill conveyor and de-energizing relays R3 and R4. The Unload conveyor will stay energized as it has a retaining contact Q 4 in parallel with the normal open of R 4. The unload conveyor will be stopped when the container reaches the limit switch L 4 at the end of the line.

That means that for all practical reasons, this circuit should work. What can now be added to it is indication lights, alarms etc. The choice is yours.

Please Note: The Author has used this control circuit as an indicative design to show the thought pattern when designing a control circuit. Before using this in a practical application, test the circuit and ensure you observe all safety requirements.

Chapter 6

Examples of Electrical Circuits

In this section we will have a look at some of the common circuits you are likely to encounter, as well as a few for you to have a play with. The aim of this is to get you to think outside of the normal pattern, be inventive!

Each of the circuits will have a brief explanation on how it works. What is important for you to remember is the fact that these circuits are the basic designs; you may find additional items added or even some removed.

Direct On Line Starter

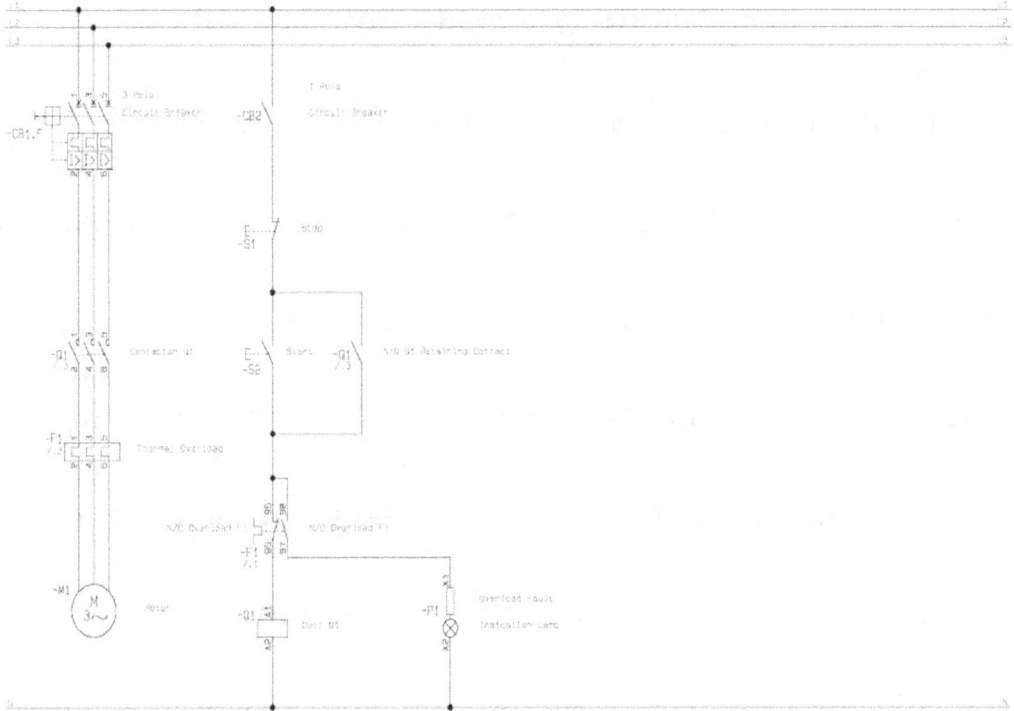

This is the most common starter you will encounter and you will see that most circuits will always contain the basic elements of this starter

Let's do the "walk-through" of the control circuit

From L1 we will go to a circuit breaker that protects our control circuit. From this circuit breaker we go to the stop button which is a normally closed contact. That means with the circuit breaker turned on, we will have power passing through the stop button up to the one side of the start and normally open contact of our contactor.

Now we push the start button which means we have power through it up to the normally closed contact of the thermal overload. If the overload is <u>not tripped,</u> we will have power through this contact to the one side of the coil and the other side of the coil is connected to Neutral. This means the coil is now energized and that will mean that the normally open contact, which is connected in parallel to the start button, will now change state and close. That means we can release the start button and the circuit will remain energized.

This contact arrangement is referred to as a Retaining Contact

Three Phase Forward Reverse

Control Circuit

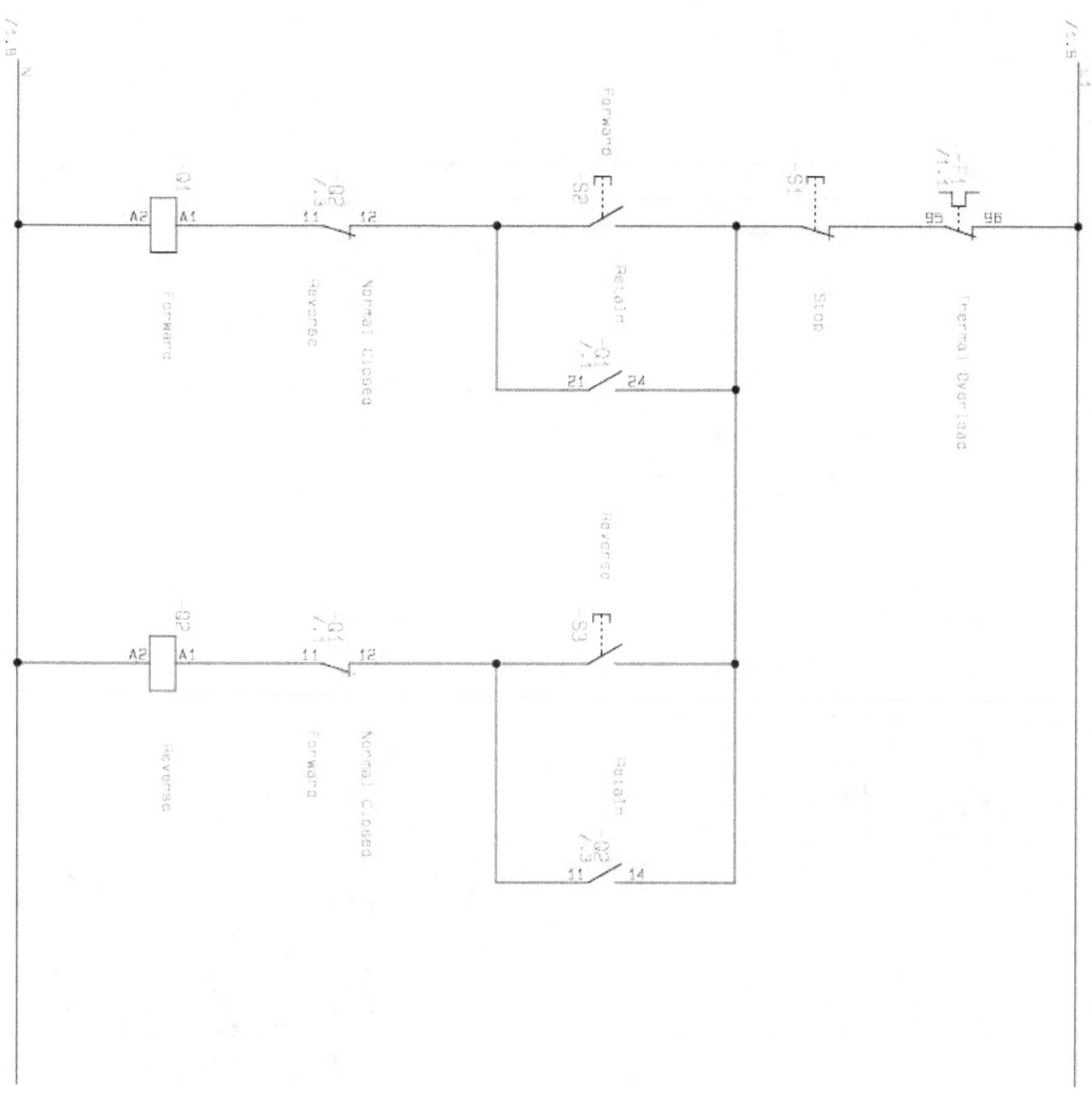

Power Circuit

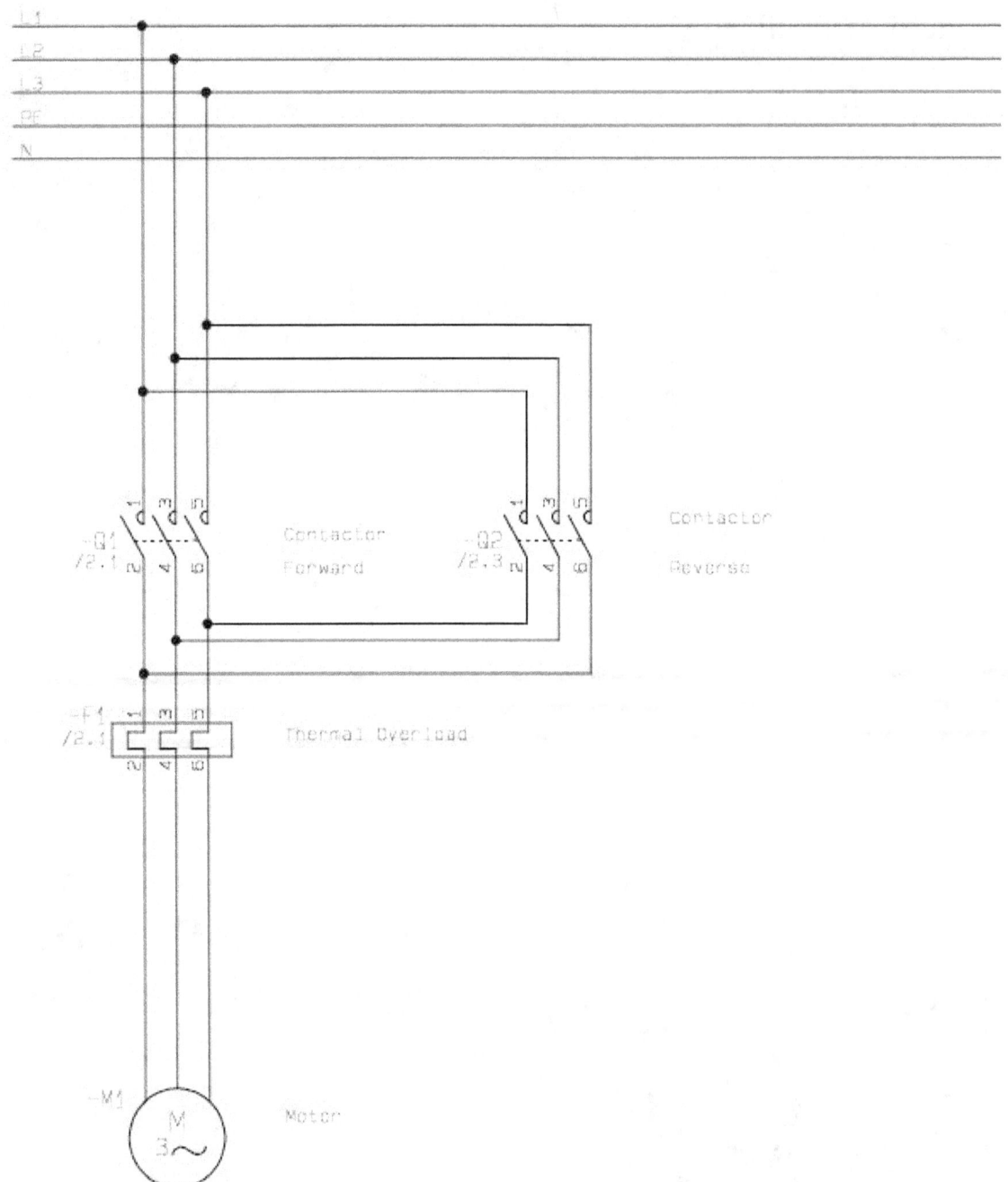

If we look at the power circuit first, you will note that the top of the contactors are connected to the same phase respectively but on the load side we swapped the red and blue phases around.

This is exactly what you would do to change direction of the motor under normal conditions i.e. Swap any two phases.

This means that when in the forward condition, you have the motor connected red, white and blue. When the motor is started in reverse, the connection changes to blue, white and red.

In the case of the control circuit, it remains very similar to the direct on line starter but we have added an interlocking contact into each section.

When you push the Forward button, the Reverse contactor has to be de-energized first. This is done by having a normal closed contact of the reverse contactor in the forward circuit. The same applies to the reverse mode. The forward contactor must be de-energized first and again we do this by placing a normally closed contact of the forward contactor in the Reverse circuit

Star/Delta

Power Circuit

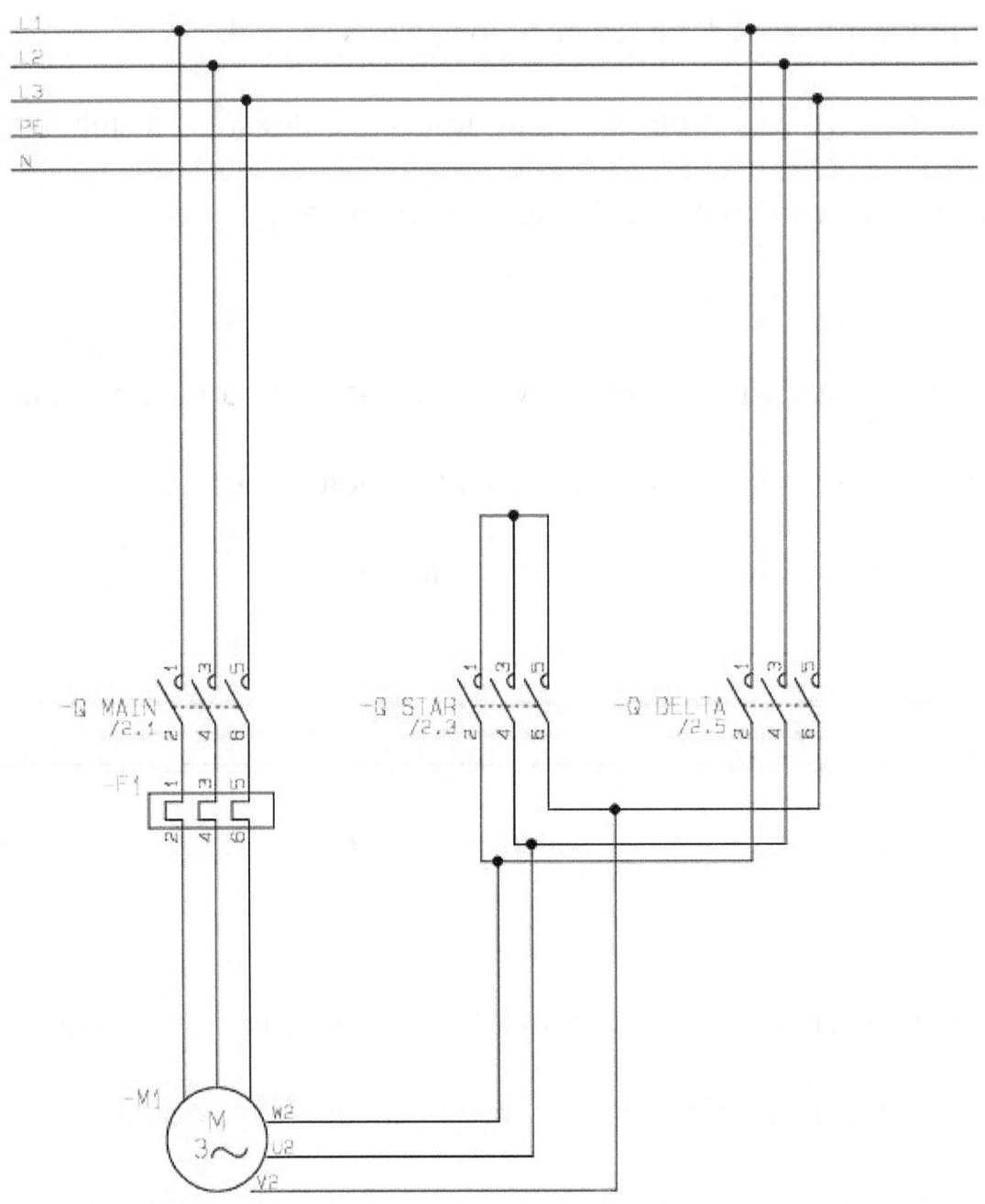

Control Circuit

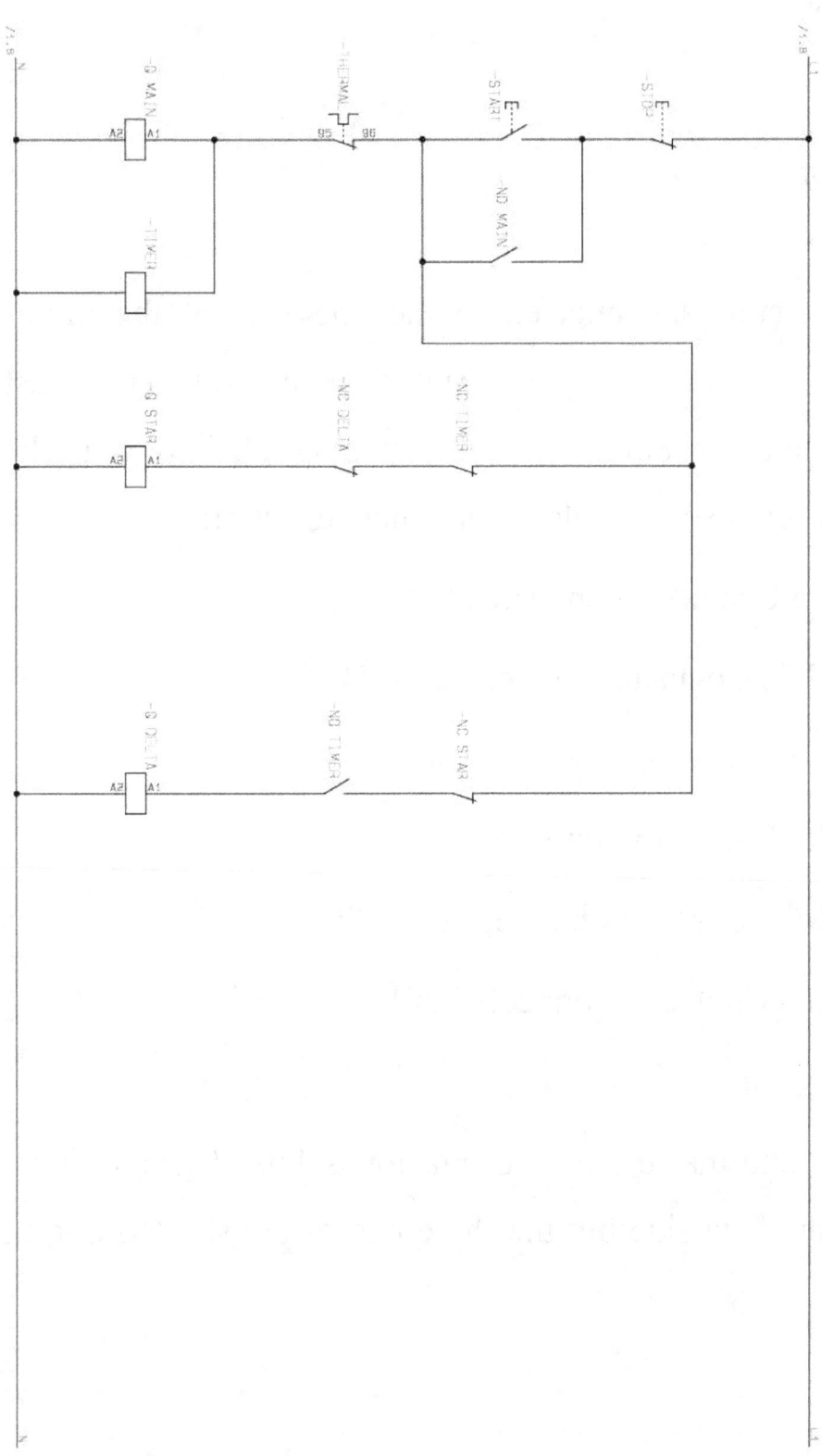

We will look at the power circuit again and this time you will note that we have three contactors in the circuit and they are:

Main Contactor

Star Contactor

Delta Contactor

The main contactor is connected to the one side of the motor windings and this is where you need to be careful. You must connect the motor windings to the matching phases from the contactors. We will use a simple method here to assist:

Red Phase Main Contactor connected to U1

White Phase Main Contactor connected to V1

Blue Phase Main Contactor connected to W1

Red Phase Delta Contactor connected to W2

White Phase Delta Contactor connected to U2

Blue Phase Delta Contactor connected to V2

You will also note that the Star Contactor is linked to the Delta Contactor on the load side but the three phases are shorted out on the line side.

That means that with the Star Contactor energized, we will have W2, U2 and V2 shorted out which is exactly what you would do if you connect a motor in the star condition

When we switch over to the Delta Contactor, we simply complete the link from U1 to W2, V1 to U2 and W1 to V2 as you would have done on the motor if you connected it in delta condition.

Now move to the control circuit. When we push the Start button, we will energize the Main Contactor which also energizes the Timer which is connected in parallel to it. At the same time we will close the retaining contact, and through the normally closed of the timer as well as through a normally closed Delta, we energize the Star Contactor. Our motor is now running in the star condition

Once the timer times out, the contacts of the timer changes state which means the normally closed contact opens and the normally open contact closes.

This makes two things happen:

The contact that now opens causes the Star Contactor to de-energize

The contact that has now closed, means that we have power to the normal closed contact of the Star, which has now closed again as the Star Contactor has de-energized. This now means that the Delta

Contactor becomes energized and our motor is running in the Delta Condition

You will also note that the power to the Star and Delta contacts is taken from below the start and retaining contact section. Should the Main Contactor become de-energized for whatever reason, we will lose power to the rest of the circuit.

Single Phase Forward/Reverse

Power Circuit

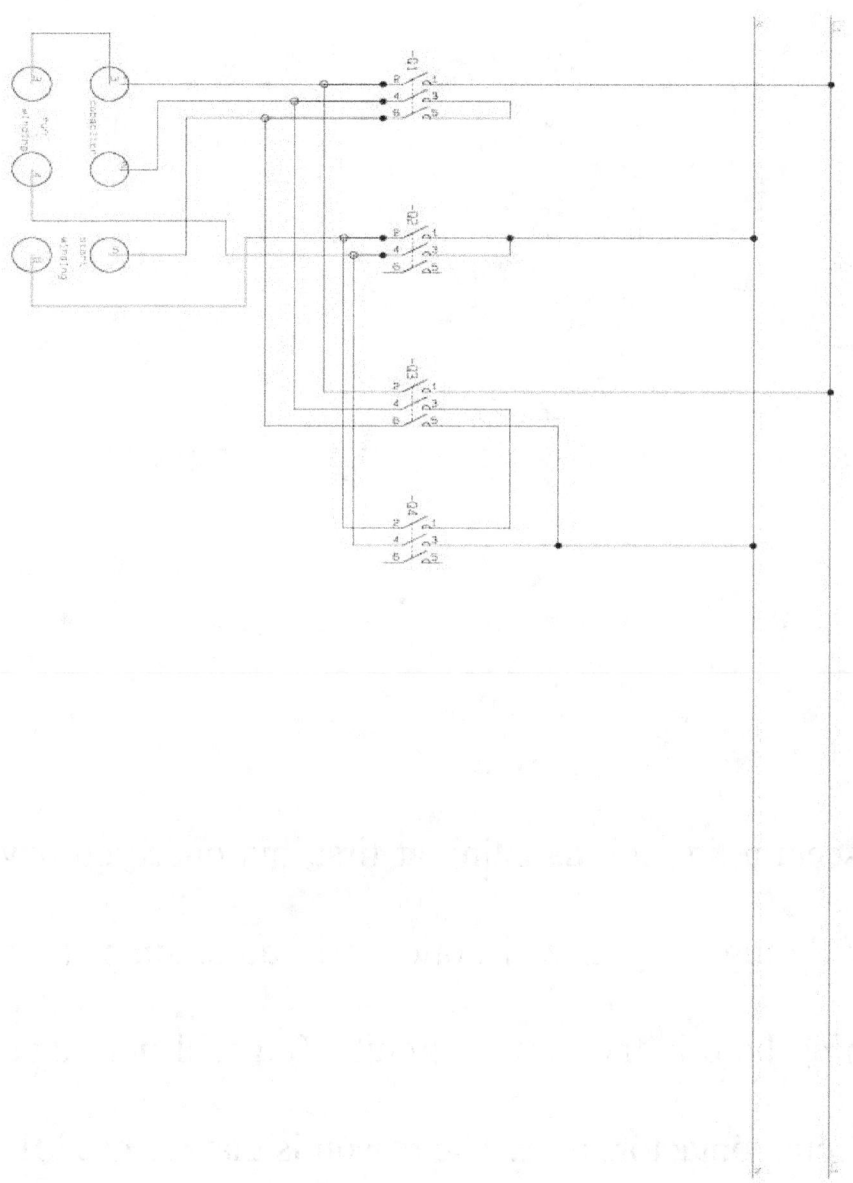

Control Circuit

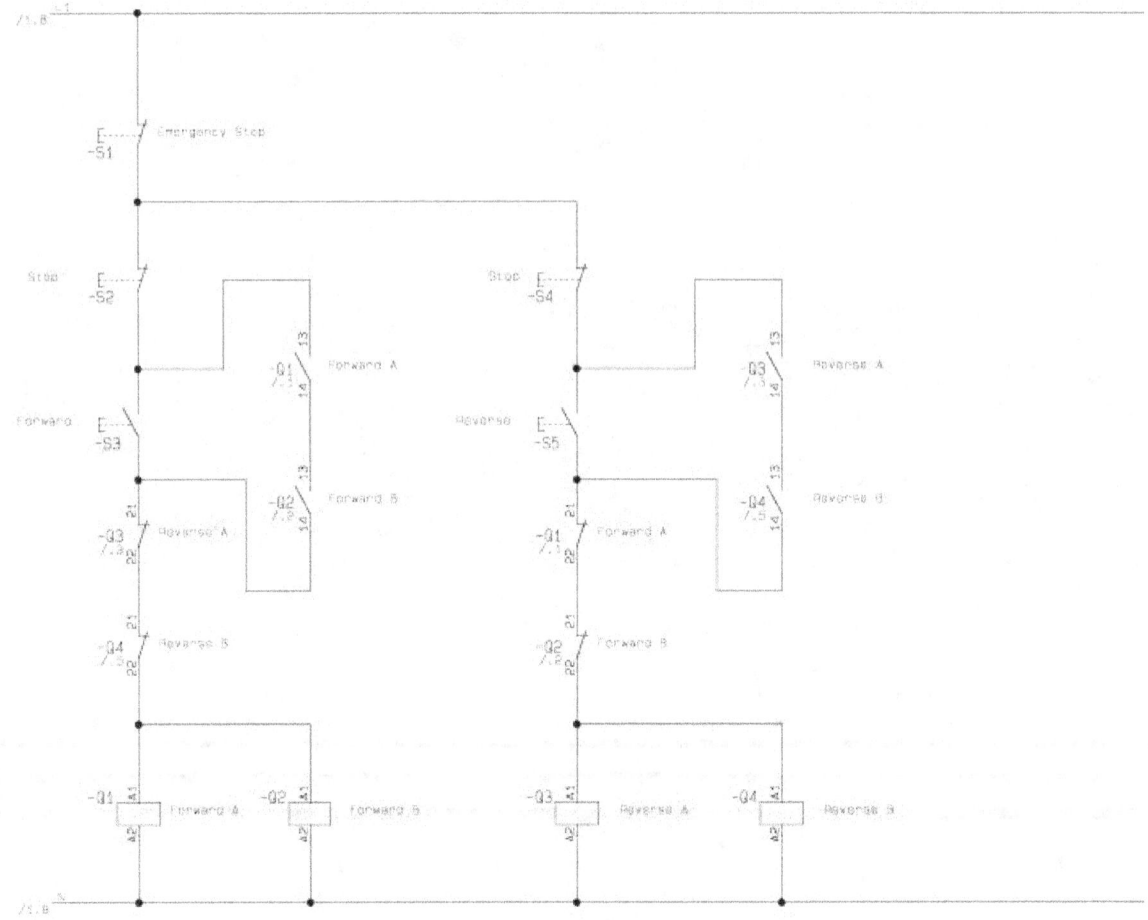

This circuit may seem a little bit daunting at first, but once you have worked your way through it you will notice it is quite simple. The most important thing here is the power circuit. You will note that I have made use of four contactors here. The reason is that we need five switching contacts for this circuit.

Following the power circuit in the Forward mode, you will see the following:

We have the Active/Live coming in on terminal 1 of the first contactor. When activated/energized, this will provide power to one side of the capacitor as well as one side of the running winding. Terminals 2 and 4 are connected to the other side of the capacitor and one side of the start winding respectively. Note the bridge across terminals 3 and 5

On contactor Q2, we have the Neutral on terminal 1 and 3. That then closes the loop to the other side of both the start and run windings when activated. The motor now runs in the forward direction. For us to change the direction, we will change the start winding around as follows:

We have again connected the Active/Live to terminal 1 of the first contactor for the reverse side, but the bridge we had from terminal 3 to 5 has now changed. We bridge from terminal 3 of Q3 to terminal 1 of Q4. Also note that our Neutral now comes in on terminal 3 of Q4 and is bridged to terminal 5 of Q3.

If you follow the motor connections now, you will note that we have swapped the start winding around with the rest of the connections still the same. Your motor now runs in reverse

For the control circuit, it is still the same principal as for the three phase forward reverse apart from the fact that I use two interlocking contacts from both the forward and reverse contactors and you will also note that I have used two contacts in series for the retaining contacts in parallel to the start buttons. Reason for this is that I want to be sure both contactors are activated.

Chapter 7

Safe Working Practices

This chapter will be based on a general overview of Safety and must be read in conjunction with any specific Standards and/or Acts pertaining to the country you are working in. There will be certain steps shown in some actions and again, please ensure that you do not substitute any Legislation with these procedures. You must at all times abide by Legislation as stipulated in the country you work in.

1 Lock Out Tag Out Procedure (LOTO)

This procedure is of utmost importance and has been made mandatory in order to protect you and/or others from serious injury. Please ensure you always follow the correct procedure and even though you have locked equipment out, test before you touch!

- Inform the relevant Supervisor/Manager that you will be working on the equipment

- Isolate the equipment and lock with an approved safety lock.

- Ensure you place your Tag on the lock. This Tag must contain the words "Do Not Operate"

- Print your name, date on which the equipment was locked out and the expected completion date on the Tag. Something I would like to see on the Tag is a contact number although this is not a requirement

- Should you not be the only person working on the equipment, make use of a hasp style clamp that has provision for multiple locks

- Once you have locked the equipment out, proceed to test all active conductors for power. Take special note of control circuits as they may have a different supply. NEVER ASSUME that it is turned off and safe to work on!

- When testing, test your meter before and after testing to make sure it is working. You may think this is a waste of time but you are relying on your meter

- Once you have completed your work, inform the Supervisor/Manager that you are ready to restore power to the equipment

- Make sure it is safe to do so and proceed to remove your lock from the equipment.

- Once you have restored power, test the equipment to ensure it is in working order before you leave.

2 Who May Remove Your Lock/Tag?

The person that has locked out the equipment must remove his/her lock and tag. Never hand your key to another person to go and remove your lock and tag from the equipment.

2 When May Another Person Remove Your Lock/Tag?

Should a scenario arise where it is necessary for your lock and tag to be removed by someone else, there are specific procedures that need to be followed.

The only person with authority to do so would normally be the Maintenance Manager and he/she must follow the following steps:

- Contact the person that has locked the equipment and verify that the work has been completed and it is safe to restore power to the equipment. The time and date of this conversation must be recorded.

- Conduct a visual inspection on the equipment to ensure all guarding has been replaced and it is safe to restore power.

- Inform the relevant Supervisor/Manager that power will be restored

- Proceed to remove lock and tag and restore power.

Although this is something that should be avoided at all costs, it does happen that a person may have left site for whatever reason without removing his/her lock and tag from the equipment. There must be a well written procedure in place for this and all staff must be trained in this procedure. Please consult your local Authority before putting this procedure in place.

3 Personal Protective Equipment

As with the above, each country and/or state may have different Legislative requirements. Please consult the relevant Standards on these and put measures in place to ensure compliance.

This does not mean that we only use it to comply with the law. These pieces of equipment are designed to protect you from serious harm or injury.

Again, there are different Standards in different countries, please ensure you check with your Local Authority as to the Standard in your place of work. In saying that, there are procedures that you need to follow and these may include:

Arc Flash Requirements, Safe Work Method Statements, Job Safety Analysis and Toolbox Meetings.

What is important is the hierarchy of controls. We need to start at the top and work our way down the list. So let's have a look at this process.

Eliminate

Is it possible to remove the hazard completely? If not, move to the next heading.

Substitute

Replace the current process/equipment to remove the risk/hazard. If this is not possible, move to the next heading

Engineer

Can we make use of engineering methods to remove the hazard? If not, move to the next heading

Isolate

In this case we must put in place measures that will separate the person from the hazard. If this is not possible, move to the next heading

Administrative Controls

This is where there needs to be formal procedures, training etc.

Personal Protective Equipment

This is virtually the last line of defense. We have not been able to control the risk in any other way. This should never be the preferred method!

Let us now take a brief look at the Arc Flash Requirements which is mandatory in some countries. This Legislation makes the PPE's mandatory when opening a cubicle in which the power is on. These would include approved clothing, face shield and "hot work" gloves. So is that enough? Certainly not. What needs to be part of this is the training and specific procedures.

Again, these documents have different names. For the purpose of this manual, I will refer to one crucial document as a Safe Work Method Statement. This document is drafted for the specific work to be undertaken and each step is clearly defined and all risks/hazards are named as well as the methods of control.

Once this document has been prepared, each step will be discussed with the personnel that are going to perform the work and every member of the team will sign the document. This document does not eliminate the risk, but makes everyone aware of each possible risk as well as how it is to be controlled. The steps in this document need to be in a logical order and this is how the task is performed.

Assume you are going to work on high voltage switchgear and the procedure for this type of work stipulates that two people will suit up with the protective equipment, a prescribed checklist is to be completed and that there will be a spotter at a safe distance.

Each of these items will be documented in the SWM and followed as per the procedure. The spotter needs to be trained in this role but apart from just watching you perform the task, he/she will also make sure you are following the SWM and Procedure. If at any point you should try to skip a step, this person will stop you immediately.

The aim of this is not to make the job harder to do; it is put in place to ensure you are kept safe. If you work at a facility where these safety measures are not in place, I urge you to put them in place. You will be astounded to see how many Electricians get very serious injuries from accidents that could have been prevented

Working at Heights

This is a topic that often sparks a debate with us. "How are we expected to do our work if we are not allowed to use a stepladder" is a common saying. We need to put this into perspective. There are way too many injuries caused from falling off a ladder and this is what prompted a big drive to try and eliminate them from a site.

Unfortunately, a lot of these accidents are caused by the person not using the ladder in a safe manner. Instead of getting down from it and move it along to the next position, it seems to be quicker to just stand right on top of it and lean over to reach right?

This is when accidents happen. The ladder may not be suited to the task but we still tend to use it because accidents only happen to other people so I should be okay?

Bottom line is, if you can avoid using a ladder, do so. Elevated work platforms are a far safer method of working at heights. Should you opt to use scaffolding, check the Standard as you may only be able to erect a scaffold that does not exceed a specific height. Anything higher that the prescribed height requires a suitably qualified person to erect it. These Standards may differ between countries so please check with your relevant Authority.

Electrical Test Equipment

These pieces of equipment are the most important part of your toolkit, yet you will not believe how some people treat them. Please look after it, your life depends on it!

1 - Your meters and test equipment need to be checked and tested before and during use to ensure they will tell you if there is power present or not.

2 The first thing to do is check the meter leads for damage or bare spots. A bare spot on the positive lead especially will cause a short to ground if connected to the power source.

3 Make sure the meter leads are rated for the power available. You can get class 3 or class 4 leads for most meters. If your meter can only read to 600 volts either lead is OK. Some meters will read to 1000 volts but can be shipped with 600 volt leads. Always make sure you know what your voltage is.

4 If you are using test equipment such as an oscilloscope that requires power to operate it should be powered from an isolation transformer.

5 Always make sure you know what you are checking for and that your meter is set to the proper scale. You do not want to be measuring 600 VAC with the meter set to ohms or amps. As well you need to know if it is AC or DC you are measuring.

6 A lot of people like to use the Tic Tracer to check for power. Please use one only for a quick check to see if there is power. When it comes to isolation to perform maintenance, you need to use a digital or analog volt meter so you can see what the voltage level is.

Summary

We have now looked at some structured approaches to working safely that are normally enforced by Legislation. What we need to look at next is how we approach our work.

You are faced with a breakdown on a machine and the pressure is on. You need to get production going and decide to work on the equipment without isolating it to try and save time. What are you thinking? One mistake and production is going to be down for a very long time and if you are lucky, you might survive it. Is that worth it? There simply is no justification to taking these risks.

In fact, any Manager worthy of his/her position will terminate your services on the spot for taking these risks in the workplace! Please always think before you do.

Another aspect of our trade is the safety of others. Never take shortcuts that could lead to disaster for the end user. If an appliance keeps on tripping the Earth Leakage (RCD) unit, there is a reason for it. I have seen an Electrician disconnect the earth wire on the appliance and turn the power back on to prove that it is not faulty because once he has done that, it no longer trips?

We may laugh about these stories but it is something that could end in tragedy. Please don't become part of the statistics! Work safe, maintain quality and above all, be proud of your work!